AUTOMATIC TRANSMISSION AND TRANSAXLE

Taskshett Manual for NATEF Proficiency

CDX Automotive

JONES & BARTLETT
LEARNING

Jones & Bartlett Learning
World Headquarters
40 Tall Pine Drive
Sudbury, MA 01776
978-443-5000
info@jblearning.com
www.jblearning.com

Jones & Bartlett Learning Canada
6339 Ormindale Way
Mississauga, Ontario L5V 1J2
Canada

Jones & Bartlett Learning International
Barb House, Barb Mews
London W6 7PA
United Kingdom

Jones & Bartlett Learning books and products are available through most bookstores and online booksellers. To contact Jones & Bartlett Learning directly, call 800-832-0034, fax 978-443-8000, or visit our website, www.jblearning.com.

Production Credits

Chairman, Board of Directors: Clayton Jones
Chief Executive Officer: Ty Field
President: James Homer
Sr. V.P., Chief Operating Officer: Don W. Jones, Jr.
V.P., Design and Production: Anne Spencer
V.P., Manufacturing and Inventory Control: Therese Connell
Publisher: Kimberly Brophy
Acquisitions Editor—Automotive: Martin Schumacher
Editorial Assistant: Kara Ebrahim
Production Manager: Jenny L. Corriveau
Director of Marketing: Alisha Weisman
Associate Marketing Manager: Jessica Cormier
Cover Design: Scott Moden
Cover Image: © Mike Zakharov/Dreamstime.com
Composition: Shepherd, Inc.
Printing and Binding: Courier Stoughton
Cover Printing: Courier Stoughton

ISBN: 978-0-7637-8499-7

The procedures and protocols in this book are based on the most current recommendations of responsible sources. The publisher, however, makes no guarantee as to, and assume no responsibility for, the correctness, sufficiency, or completeness of such information or recommendations. Other or additional safety measures may be required under particular circumstances.

For every task in ASE 2: Automatic Transmission and Transaxle, the following safety requirement must be strictly followed: Comply with personal and environmental safety practices associated with clothing; eye protection; hand tools; power equipment; proper ventilation; and the handling, storage, and disposal of chemicals/materials in accordance with local, state, and federal safety and environmental regulations.

6048
Printed in the United States of America
14 13 12 11 10 10 9 8 7 6 5 4 3 2

Contents

Resource Preview

2008 NATEF and CDX tasksheet numbers appear at the beginning and end of every task, as well as in the general information for the section.

Required and recommended materials and equipment are listed for each task.

Safety issues relevant to the tasks are listed at the beginning of every section.

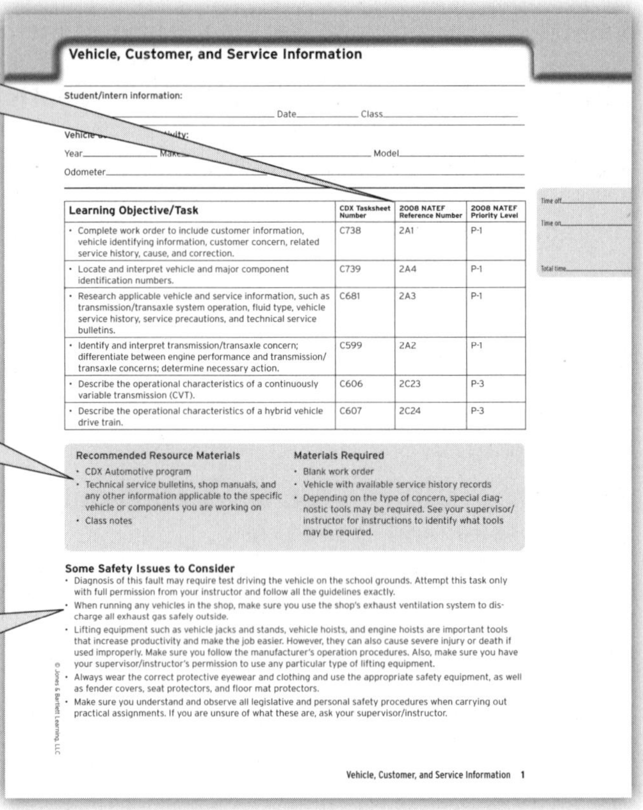

Vehicle, Customer, and Service Information

Student/intern information: Date Class

Year Make Model

Odometer

Time off / Time on / Total time

Learning Objective/Task	CDX Tasksheet Number	2008 NATEF Reference Number	2008 NATEF Priority Level
• Complete work order to include customer information, vehicle identifying information, customer concern, related service history, cause, and correction.	C738	2A1	P-1
• Locate and interpret vehicle and major component identification numbers.	C739	2A4	P-1
• Research applicable vehicle and service information, such as transmission/transaxle system operation, fluid type, vehicle service history, service precautions, and technical service bulletins.	C681	2A3	P-1
• Identify and interpret transmission/transaxle concern; differentiate between engine performance and transmission/transaxle concerns; determine necessary action.	C599	2A2	P-1
• Describe the operational characteristics of a continuously variable transmission (CVT).	C606	2C23	P-3
• Describe the operational characteristics of a hybrid vehicle drive train.	C607	2C24	P-3

Recommended Resource Materials
- CDX Automotive program
- Technical service bulletins, shop manuals, and any other information applicable to the specific vehicle or components you are working on
- Class notes

Materials Required
- Blank work order
- Vehicle with available service history records
- Depending on the type of concern, special diagnostic tools may be required. See your supervisor/instructor for instructions to identify what tools may be required.

Some Safety Issues to Consider
- Diagnosis of this fault may require test driving the vehicle on the school grounds. Attempt this task only with full permission from your instructor and follow all the guidelines exactly.
- When running any vehicles in the shop, make sure you use the shop's exhaust ventilation system to discharge all exhaust gas safely outside.
- Lifting equipment such as vehicle jacks and stands, vehicle hoists, and engine hoists are important tools that increase productivity and make the job easier. However, they can also cause severe injury or death if used improperly. Make sure you follow the manufacturer's operation procedures. Also, make sure you have your supervisor/instructor's permission to use any particular type of lifting equipment.
- Always wear the correct protective eyewear and clothing and use the appropriate safety equipment, as well as fender covers, seat protectors, and floor mat protectors.
- Make sure you understand and observe all legislative and personal safety procedures when carrying out practical assignments. If you are unsure of what these are, ask your supervisor/instructor.

© Jones & Bartlett Learning, LLC

Vehicle, Customer, and Service Information 1

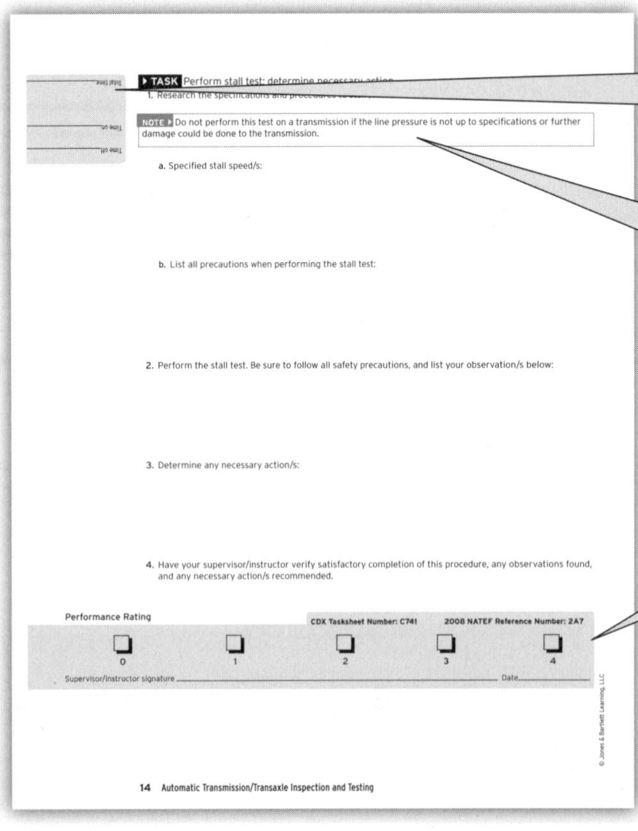

Total time / Time on / Time off

TASK Perform stall test; determine necessary action.

1. Research the specifications and procedures

NOTE Do not perform this test on a transmission if the line pressure is not up to specifications or further damage could be done to the transmission.

a. Specified stall speed/s:

b. List all precautions when performing the stall test:

2. Perform the stall test. Be sure to follow all safety precautions, and list your observation/s below:

3. Determine any necessary action/s:

4. Have your supervisor/instructor verify satisfactory completion of this procedure, any observations found, and any necessary action/s recommended.

Performance Rating — CDX Tasksheet Number: C741 — 2008 NATEF Reference Number: 2A7

0 1 2 3 4

Supervisor/instructor signature Date

© Jones & Bartlett Learning, LLC

14 Automatic Transmission/Transaxle Inspection and Testing

Time card feature appears at the beginning of every task and allows students to track the time they spend on each task.

Note boxes offer students vital information to consider while performing the task.

Performance standard and rating areas simplify the sign-off process for instructors.

Acknowledgments

Jones & Bartlett Learning would like to thank the following contributor and reviewers for their assistance with this manual:

Contributor:

Kirk VanGelder, CMAT & L1, NATEF ETL
AYES Instructor
Clark County Skills Center
Vancouver, Washington

Reviewers:

Edward J. Heim
ASE certified in Automotive and Automotive Machinist
Battle Ground High School
Battle Ground, Washington

Marty Kamimoto
Fresno City College
Fresno, California

Sonny Reeves, CMAT, NATEF ETL
AYES Instructor
Hutchings Career Center
Macon, Georgia

Jim Stafford
Tennessee Technology Center—Newbern
Newbern, Tennessee

CDX Automotive

Designed to help automotive programs meet NATEF requirements, this interactive automotive training system introduces students to the fundamental principles and applications of automotive education. It also equips instructors with a complete set of easy-to-use course materials. This innovative resource will enrich your conventional course, or can be used as a primary management source to deliver an online program that accommodates increasing student enrollments.

CDX Automotive improves learning outcomes in automotive education by increasing student engagement and performance through personalized learning. Key features include:

- Course content mapped to the 8 ASE test areas
- Easy and intuitive navigation
- Over 850 automotive procedural and demonstration videos, including content on timely topics such as alternative fuels
- Unlimited practice tests, knowledge check exercises, and final exams that are automatically scored and recorded to save time and eliminate human error
- Support for both 2005 and 2008 NATEF standards with tasksheets and updated tasksheet reporting, making ASE recertification even easier for instructors nationwide
- *CDX Automotive Access Pack* option for students looking to purchase required or recommended access to ***CDX Automotive*** from their local college bookstore

CDX Automotive includes easy-to-use and state-of-the-art learning performance management tools to help schools manage and report student performance data that meet NATEF's ASE e-learning certification requirements. ***CDX Automotive*** offers a turnkey solution to high schools, vocational schools, community colleges, and skill centers wishing to join the e-learning initiative announced by NATEF that allows programs to qualify for reduced curriculum hour requirements through e-learning completed outside of regular program hours.

Steer your course in the direction you want with ***CDX Automotive***.

Request a free trial at www.cdxauto.com.

Vehicle, Customer, and Service Information

Student/intern information:

Name__________ Date__________ Class__________

Vehicle used for this activity:

Year__________ Make__________ Model__________

Odometer__________ VIN __________

Time off__________

Time on__________

Total time__________

Learning Objective/Task	CDX Tasksheet Number	2008 NATEF Reference Number	2008 NATEF Priority Level
• Complete work order to include customer information, vehicle identifying information, customer concern, related service history, cause, and correction.	C738	2A1	P-1
• Locate and interpret vehicle and major component identification numbers.	C739	2A4	P-1
• Research applicable vehicle and service information, such as transmission/transaxle system operation, fluid type, vehicle service history, service precautions, and technical service bulletins.	C681	2A3	P-1
• Identify and interpret transmission/transaxle concern; differentiate between engine performance and transmission/transaxle concerns; determine necessary action.	C599	2A2	P-1
• Describe the operational characteristics of a continuously variable transmission (CVT).	C606	2C23	P-3
• Describe the operational characteristics of a hybrid vehicle drive train.	C607	2C24	P-3

Recommended Resource Materials

- CDX Automotive program
- Technical service bulletins, shop manuals, and any other information applicable to the specific vehicle or components you are working on
- Class notes

Materials Required

- Blank work order
- Vehicle with available service history records
- Depending on the type of concern, special diagnostic tools may be required. See your supervisor/instructor for instructions to identify what tools may be required.

Some Safety Issues to Consider

- Diagnosis of this fault may require test driving the vehicle on the school grounds. Attempt this task only with full permission from your instructor and follow all the guidelines exactly.
- When running any vehicles in the shop, make sure you use the shop's exhaust ventilation system to discharge all exhaust gas safely outside.
- Lifting equipment such as vehicle jacks and stands, vehicle hoists, and engine hoists are important tools that increase productivity and make the job easier. However, they can also cause severe injury or death if used improperly. Make sure you follow the manufacturer's operation procedures. Also, make sure you have your supervisor/instructor's permission to use any particular type of lifting equipment.
- Always wear the correct protective eyewear and clothing and use the appropriate safety equipment, as well as fender covers, seat protectors, and floor mat protectors.
- Make sure you understand and observe all legislative and personal safety procedures when carrying out practical assignments. If you are unsure of what these are, ask your supervisor/instructor.

Performance Standard

0–No exposure: No information or practice provided during the program; complete training required
1–Exposure only: General information provided with no practice time; close supervision needed; additional training required
2–Limited practice: Has practiced job during training program; additional training required to develop skill
3–Moderately skilled: Has performed job independently during training program; limited additional training may be required
4–Skilled: Can perform job independently with no additional training

Total time________________
Time on________________
Time off________________

▸ TASK Complete work order to include customer information, vehicle identifying information, customer concern, related service history, cause, and correction. **C738 2A1**

1. Complete the work order, specifying the following:
   a. Customer information
   b. Customer concern
   c. Vehicle identifying information
   d. Any related service history, etc.

The rest of this task is completed by performing the next three tasks, and can be signed off at the end of this tasksheet, on page 5.

Total time________________
Time on________________
Time off________________

▸ TASK Locate and interpret vehicle and major component identification numbers. **C739 2A4**

1. Using the vehicle VIN for identification, access the appropriate technical information source to identify the following details.
   a. VIN: ____
2. Look up the VIN in the appropriate service manual and identify the following information.
   a. Country of origin: ________________________________
   b. Plant: ________________________________
   c. Type of restraint system: ________________________________
   d. Engine: ________________________________
   e. Model year: ________________________________
3. List the manufacturer's designation for this transmission: ________________________________
4. List the location of any certification and calibration decals:

5. Describe the main information on each decal:

6. Have your supervisor/instructor verify satisfactory completion of this procedure, any observations found, and any necessary action/s recommended.

Performance Rating

CDX Tasksheet Number: C739 **2008 NATEF Reference Number: 2A4**

☐ 0 ☐ 1 ☐ 2 ☐ 3 ☐ 4

Supervisor/instructor signature ______________________ Date______

▶ TASK Research applicable vehicle and service information, such as transmission/transaxle system operation, fluid type, vehicle service history, service precautions, and technical service bulletins. C681 2A3

Time off______

Time on______

Total time______

1. Using the vehicle VIN for identification, use the appropriate source to access the vehicle's service history in relation to prior related transmission/transaxle work or customer concerns.
   a. List any related repairs/concerns, and their dates:

2. List the specified type of fluid for this vehicle: ______________________
3. Using the vehicle VIN for identification, access any relevant technical service bulletins for the particular vehicle you are working on in relation to any transmission/transaxle updates, precautions, or other service issues. List any related service bulletins:

4. Have your supervisor/instructor verify satisfactory completion of this procedure, any observations found, and any necessary action/s recommended.

Performance Rating

CDX Tasksheet Number: C681 **2008 NATEF Reference Number: 2A3**

☐ 0 ☐ 1 ☐ 2 ☐ 3 ☐ 4

Supervisor/instructor signature ______________________ Date______

Total time______

Time on______

Time off______

▶ TASK Identify and interpret transmission/transaxle concern; differentiate between engine performance and transmission/transaxle concerns; determine necessary action. **C599 2A2**

1. List the customer concern/s:

2. Research the particular concern in the appropriate service manual.
   a. List the possible causes:

3. Inspect the vehicle to determine the cause of the concern (differentiate between engine performance and transmission/transaxle concerns).
   a. List the steps you took to determine the fault/s:

4. List the cause of the concern/s:

5. List the necessary action/s to correct the fault/s:

6. Have your supervisor/instructor verify satisfactory completion of this procedure, any observations found, and any necessary action/s recommended.

Performance Rating

CDX Tasksheet Number: C599   2008 NATEF Reference Number: 2A2

☐ 0   ☐ 1   ☐ 2   ☐ 3   ☐ 4

Supervisor/instructor signature ______________________ Date__________

TASK C738/2A1 Continued Complete work order to include customer information, vehicle identifying information, customer concern, related service history, cause, and correction.

Time off__________

Time on__________

Total time__________

NOTE The following sign off goes along with the first task on this tasksheet and can also be signed off once the previous three tasks in this tasksheet are completed.

2. Have your supervisor/instructor verify satisfactory completion of this procedure, any observations found, and any necessary action/s recommended.

Performance Rating

CDX Tasksheet Number: C738   2008 NATEF Reference Number: 2A1

☐ 0   ☐ 1   ☐ 2   ☐ 3   ☐ 4

Supervisor/instructor signature ______________________ Date__________

TASK Describe the operational characteristics of a continuously variable transmission (CVT).

C606 2C23

Vehicle used for this activity:

Year__________ Make__________ Model__________

Odometer__________ VIN__________

Time off__________

Time on__________

Total time__________

1. Research the description and operation of a continuously variable transmission in the appropriate service information.
   a. What type of CVT is this transmission?

b. How does this CVT achieve reduction?

c. How does this CVT achieve direct drive?

d. How does this CVT achieve neutral?

e. How does this CVT achieve reverse?

f. What type of fluid does this CVT use? ______________________

g. What is the fluid capacity? ______________________ qt/lt

2. Have your supervisor/instructor verify satisfactory completion of this procedure, any observations found, and any necessary action/s recommended.

Performance Rating

CDX Tasksheet Number: C606 **2008 NATEF Reference Number: 2C23**

☐ 0 ☐ 1 ☐ 2 ☐ 3 ☐ 4

Supervisor/instructor signature ______________________ Date______

Time off ______

Time on ______

Total time ______

▶ TASK Describe the operational characteristics of a hybrid vehicle drive train. **C607 2C24**

1. Research the description and operation of a hybrid vehicle drive train in the appropriate service information.
   a. What type of hybrid is this vehicle? **Series/Parallel/Series-Parallel** (Circle one)
   b. Can this vehicle operate only on the internal combustion engine? **Yes/No** (Circle one)

c. Describe the conditions when the internal combustion engine runs:

d. Can this vehicle operate only on the electric motor? **Yes/No** (Circle one)

e. Describe the conditions when the electric motor operates:

f. Does this vehicle use "regeneration"? **Yes/No** (Circle one)

g. Describe how "regeneration" operates in this vehicle:

2. Have your supervisor/instructor verify satisfactory completion of this procedure, any observations found, and any necessary action/s recommended.

Performance Rating

CDX Tasksheet Number: C607 **2008 NATEF Reference Number: 2C24**

☐	☐	☐	☐	☐
0	1	2	3	4

Supervisor/instructor signature ______________________________ Date____________

Student/intern information:

Name________________________________ Date____________ Class______________________

Recommended Resource Materials	Materials Required
• CDX Automotive program • Technical service bulletins, shop manuals, and any other information applicable to the specific vehicle or components you are working on • Class notes	• None, but your instructor may want you to use a copy of the school's repair order to write up the 3 Cs

Some Safety Issues to Consider

- Always wear the correct protective eyewear and clothing and use the appropriate safety equipment, as well as fender covers, seat protectors, and floor mat protectors.
- Make sure you understand and observe all legislative and personal safety procedures when carrying out practical assignments. If you are unsure of what these are, ask your supervisor/instructor.

Performance Standard

0–No exposure: No information or practice provided during the program; complete training required
1–Exposure only: General information provided with no practice time; close supervision needed; additional training required
2–Limited practice: Has practiced job during training program; additional training required to develop skill
3–Moderately skilled: Has performed job independently during training program; limited additional training may be required
4–Skilled: Can perform job independently with no additional training

▶ PRACTICE TASK Demonstrate use of 3 Cs (concern, cause, and correction).

1. Using the following scenario, write up the 3 Cs as listed on most repair orders. Assume that the customer authorized the recommended repairs.

   A compact vehicle is brought to your shop with an automatic transaxle concern. The customer tells you that the transmission started slipping a couple of days ago, and it is getting worse. You road test the vehicle and notice that it is slipping in all gears when under moderate load. You pull the vehicle onto a rack, and find the following:

   a. The vehicle has 122,000 miles on it, but, otherwise, is in good condition.
   b. The transmission fluid is dark and smells burnt. The fluid is about two pints too low, but topping it off does not stop the vehicle from slipping when it is stall tested.
   c. The transmission pan gasket and both axle seals are seeping heavily.
   d. With the pan removed, there is a lot of band and clutch material in the bottom, but very few metal filings.
   e. The transmission mount is torn.
   f. Both outer CV joint boots are slightly torn, but the joints are still good.

NOTE ▶ Ask your instructor whether you should use a copy of the shop repair order or the 3 Cs below to record this information.

2. Concern:

3. Cause:

4. Correction:

5. Other recommended service:

6. Have your supervisor/instructor verify satisfactory completion of this procedure, any observations found, and any necessary action/s recommended.

Performance Rating

❑ 0 ❑ 1 ❑ 2 ❑ 3 ❑ 4

Supervisor/instructor signature ______________________________ Date__________

Automatic Transmission/Transaxle Inspection and Testing

Student/intern information:

Name____________________ Date__________ Class________________

Vehicle used for this activity:

Year__________ Make____________________ Model________________

Odometer____________________ VIN ____________________

Time off__________

Time on__________

Total time__________

Learning Objective/Task	CDX Tasksheet Number	2008 NATEF Reference Number	2008 NATEF Priority Level
• Diagnose fluid loss and condition concerns; check fluid level in transmissions with and without dip-stick; determine necessary action.	C682	2A5	P-1
• Perform pressure tests (including transmissions/transaxles equipped with electronic pressure control); determine necessary action.	C740	2A6	P-1
• Perform stall test; determine necessary action.	C741	2A7	P-3
• Perform lock-up converter system tests; determine necessary action.	C063	2A8	P-3

Recommended Resource Materials

- CDX Automotive program
- Technical service bulletins, shop manuals, and any other information applicable to the specific vehicle or components you are working on
- Class notes

Materials Required

- Vehicle with automatic transmission/transaxle and lock-up converter
- Vehicle hoist
- Flashlight or shop light
- Optional: fluorescent dye and ultraviolet light
- Transmission pressure gauge set
- Tachometer
- Scan tool
- DVOM/DMM

Some Safety Issues to Consider

- Vehicle hoists are important tools that increase productivity and make the job easier. However, they can also cause severe injury or death if used improperly. Make sure you follow the manufacturer's operation procedures. Also, make sure you have your supervisor/instructor's permission to use any particular type of lifting equipment.
- This procedure may require cleaning the transmission or axle assembly prior to inspection. If so, use the proper equipment and procedures to carry this out safely. Also, get your supervisor/instructor's permission prior to cleaning.
- **Caution**: Stall testing for more than a few seconds can damage the transmission or other parts of the vehicle. It is also a hazardous task. Perform this task with utmost care!
- When running any vehicles in the shop, make sure you use the shop's exhaust ventilation system to discharge all exhaust gas safely outside.
- Diagnosis of this fault may require test driving the vehicle on the school grounds. Attempt this task only with full permission from your instructor and follow all the guidelines exactly.
- Always wear the correct protective eyewear and clothing and use the appropriate safety equipment, as well as fender covers, seat protectors, and floor mat protectors.
- Make sure you understand and observe all legislative and personal safety procedures when carrying out practical assignments. If you are unsure of what these are, ask your supervisor/instructor.

Performance Standard

0–No exposure: No information or practice provided during the program; complete training required
1–Exposure only: General information provided with no practice time; close supervision needed; additional training required
2–Limited practice: Has practiced job during training program; additional training required to develop skill
3–Moderately skilled: Has performed job independently during training program; limited additional training may be required
4–Skilled: Can perform job independently with no additional training

Total time______________

Time on______________

Time off______________

▶ TASK Diagnose fluid loss and condition concerns; check fluid level in transmissions with and without dip-stick; determine necessary action. **C682 2A5**

1. List the transmission designation for this vehicle: ____________________________
2. Research the following specifications for this vehicle in the appropriate service information.
   a. Fluid type: ____________________________
   b. Fluid capacity: ____________________________ qt/lt
   c. Is the fluid level checked with a dip-stick? **Yes/No** (Circle one)
   d. If not, how is the fluid level checked?
3. Safely raise the vehicle on a hoist.
4. Inspect the transmission/transaxle assembly for leaks.
   a. Identify and list the source of any leak/s:
5. Check the fluid level and list your observation/s:
6. Inspect the condition of the fluid. Look for contaminants, metal shavings, and degraded fluid condition. List your observation/s:

7. Determine any necessary action/s:

8. Have your supervisor/instructor verify satisfactory completion of this procedure, any observations found, and any necessary action/s recommended.

Performance Rating

CDX Tasksheet Number: C682 **2008 NATEF Reference Number: 2A5**

☐	☐	☐	☐	☐
0	1	2	3	4

Supervisor/instructor signature ______________________________ Date______________

▸ TASK Perform pressure tests (including transmissions/transaxles equipped with electronic pressure control); determine necessary action. C740 2A6

Time off________

Time on________

Total time________

1. Research all applicable specifications and test procedure/s for this transmission in the appropriate service information.
   a. Line pressure: ______________________ psi/kPa
   b. Governor pressure: ______________________ psi/kPa
   c. Low gear servo/clutch pressure: ______________________ psi/kPa
   d. Second gear servo/clutch pressure: ______________________ psi/kPa
   e. Direct clutch pressure: ______________________ psi/kPa
   f. Cooler circuit pressure: ______________________ psi/kPa
2. Perform the applicable transmission pressure tests. Be sure to follow the specified procedures, and list your observations below.
   a. Line pressure: ______________________ psi/kPa
   b. Governor pressure: ______________________ psi/kPa
   c. Low gear servo/clutch pressure: ______________________ psi/kPa
   d. Second gear servo/clutch pressure: ______________________ psi/kPa
   e. Direct clutch pressure: ______________________ psi/kPa
   f. Cooler circuit pressure: ______________________ psi/kPa
3. Determine any necessary action/s:

4. Have your supervisor/instructor verify satisfactory completion of this procedure, any observations found, and any necessary action/s recommended.

Performance Rating

CDX Tasksheet Number: C740 **2008 NATEF Reference Number: 2A6**

☐	☐	☐	☐	☐
0	1	2	3	4

Supervisor/instructor signature ______________________________ Date______________

Total time

Time on

Time off

▶ TASK Perform stall test; determine necessary action. C741 2A7

1. Research the specifications and procedures to stall test this transmission.

NOTE ▶ Do not perform this test on a transmission if the line pressure is not up to specifications or further damage could be done to the transmission.

a. Specified stall speed/s:

b. List all precautions when performing the stall test:

2. Perform the stall test. Be sure to follow all safety precautions, and list your observation/s below:

3. Determine any necessary action/s:

4. Have your supervisor/instructor verify satisfactory completion of this procedure, any observations found, and any necessary action/s recommended.

Performance Rating

CDX Tasksheet Number: C741 **2008 NATEF Reference Number: 2A7**

☐ 0 ☐ 1 ☐ 2 ☐ 3 ☐ 4

Supervisor/instructor signature ______________________ Date__________

▶ TASK Perform lock-up converter system tests; determine necessary action. C063 2A8

Time off________

Time on________

Total time________

1. Research the specifications and procedure for testing the lock-up converter.
   a. Lock-up converter solenoid resistance: ____________________ ohm
   b. List, or print off and attach to this sheet, the procedure for testing the lock-up converter:

   c. Ask your instructor where you are to perform the lock-up converter test: **Road test/Vehicle hoist** (Circle one)
2. Have your supervisor/instructor approve your answers and initial below.
   a. Supervisor/instructor's initials: ____________________
3. Perform the lock-up converter performance test.
   a. Prepare the vehicle for this test by following the specified procedure.
   b. Test the operation of the lock-up converter system following your instructor's directions. List your observation/s:

4. Perform the lock-up converter pin-point test.
   a. Test for the proper electrical signal at the lock-up converter solenoid following the specified procedure. List your observation/s:

   b. Measure the electrical continuity/resistance of the solenoid: ____________________ ohm
5. Determine any necessary action/s:

6. Have your supervisor/instructor verify satisfactory completion of this procedure, any observations found, and any necessary action/s recommended.

Performance Rating

CDX Tasksheet Number: C063 **2008 NATEF Reference Number: 2A8**

☐ 0 ☐ 1 ☐ 2 ☐ 3 ☐ 4

Supervisor/instructor signature ____________________ Date________

Automatic Transmission/Transaxle Diagnosis

Student/intern information:

Name________________ Date________ Class__________

Vehicle used for this activity:

Year________ Make________________ Model__________

Odometer________________ VIN ________________

Time off________

Time on________

Total time________

Learning Objective/Task	CDX Tasksheet Number	2008 NATEF Reference Number	2008 NATEF Priority Level
• Diagnose electronic transmission/transaxle control systems using appropriate test equipment and service information.	C600	2A13	P-1
• Diagnose transmission/transaxle gear reduction/ multiplication concerns using driving, driven, and held member (power flow) principles.	C066	2A10	P-1
• Diagnose pressure concerns in a transmission using hydraulic principles (Pascal's Law).	C747	2A12	P-2
• Diagnose noise and vibration concerns; determine necessary action.	C743	2A9	P-2

Recommended Resource Materials

- CDX Automotive program
- Technical service bulletins, shop manuals, and any other information applicable to the specific vehicle or components you are working on
- Class notes

Materials Required

- Vehicle with automatic transmission/transaxle
- Vehicle hoist
- Transmission pressure gauge set
- Tachometer
- Scan tool
- DVOM/DMM

Some Safety Issues to Consider

- Vehicle hoists are important tools that increase productivity and make the job easier. However, they can also cause severe injury or death if used improperly. Make sure you follow the manufacturer's operation procedures. Also, make sure you have your supervisor/instructor's permission to use any particular type of lifting equipment.
- When running any vehicles in the shop, make sure you use the shop's exhaust ventilation system to discharge all exhaust gas safely outside.
- Diagnosis of this fault may require test driving the vehicle on the school grounds. Attempt this task only with full permission from your instructor and follow all the guidelines exactly.
- Always wear the correct protective eyewear and clothing and use the appropriate safety equipment, as well as fender covers, seat protectors, and floor mat protectors.
- Make sure you understand and observe all legislative and personal safety procedures when carrying out practical assignments. If you are unsure of what these are, ask your supervisor/instructor.

Performance Standard

0–No exposure: No information or practice provided during the program; complete training required
1–Exposure only: General information provided with no practice time; close supervision needed; additional training required
2–Limited practice: Has practiced job during training program; additional training required to develop skill
3–Moderately skilled: Has performed job independently during training program; limited additional training may be required
4–Skilled: Can perform job independently with no additional training

Total time________________

Time on________________

Time off________________

▶ TASK Diagnose electronic transmission/transaxle control systems using appropriate test equipment and service information.

C600 2A13

1. List the customer concern related to an electronic control system issue:

2. Connect an appropriate scan tool to the data link connector (DLC). Retrieve and list any transmission-related codes:

3. Research the specifications and procedures for diagnosing the code/s.
   a. List the possible cause/s of the code/s:

   b. List, or print off and attach to this sheet, the procedure for diagnosing the code/s:

4. Perform the diagnostic tests following the specified procedure using the appropriate test equipment.
   a. List your tests and test results:

   b. List the test equipment you used to diagnose this concern:

5. Determine any necessary action/s to correct the fault:

6. Have your supervisor/instructor verify satisfactory completion of this procedure, any observations found, and any necessary action/s recommended.

Performance Rating

CDX Tasksheet Number: C600 **2008 NATEF Reference Number: 2A13**

☐	☐	☐	☐	☐
0	1	2	3	4

Supervisor/instructor signature ____________________ Date________

▶ TASK Diagnose transmission/transaxle gear reduction/multiplication concerns using driving, driven, and held member (power flow) principles. C066 2A10

Time off________

Time on________

Total time________

1. List the customer concern:

2. Research the particular transmission/transaxle concern in the appropriate service information, including power flow charts and driving, driven, and held member charts.
   a. List the possible cause/s of the concern:

   b. List, or print off and attach to this sheet, the procedure for diagnosing the concern:

3. Perform the diagnostic tests following the specified procedure.
   a. List your tests and test results:

b. Describe how you used the diagnostic charts to diagnose the concern:

4. Determine any necessary action/s to correct the fault:

5. Have your supervisor/instructor verify satisfactory completion of this procedure, any observations found, and any necessary action/s recommended.

Performance Rating

CDX Tasksheet Number: C066 **2008 NATEF Reference Number: 2A10**

0	1	2	3	4
☐	☐	☐	☐	☐

Supervisor/instructor signature ____________________ Date________

Total time________

Time on________

Time off________

▶ TASK Diagnose pressure concerns in a transmission using hydraulic principles (Pascal's Law).

C747 2A12

1. List the transmission/transaxle pressure-related concern:

2. Research the concern in the appropriate service information.
   a. List the possible cause/s of the concern:

   b. List any related specified pressures:

3. Write out Pascal's Law:

4. Describe how you will use Pascal's Law to help you diagnose this concern:

5. Perform the appropriate pressure tests and list your observation/s:

6. Determine any necessary action/s to correct the fault:

7. Have your supervisor/instructor verify satisfactory completion of this procedure, any observations found, and any necessary action/s recommended.

Performance Rating

CDX Tasksheet Number: C747 **2008 NATEF Reference Number: 2A12**

☐ 0 ☐ 1 ☐ 2 ☐ 3 ☐ 4

Supervisor/instructor signature ______________________ Date__________

▶ TASK Diagnose noise and vibration concerns; determine necessary action. **C743 2A9**

1. List the transmission noise and vibration-related concern:

Time off__________

Time on__________

Total time__________

2. Research the concern in the appropriate service information.
   a. List the possible cause/s of the concern:

   b. List, or print off and attach to this sheet, the specified procedure for diagnosing the concern:

3. Perform the appropriate tests, following the specified procedure. List your observation/s:

4. List the cause/s of the concern:

5. Determine any necessary action/s to correct the fault:

6. Have your supervisor/instructor verify satisfactory completion of this procedure, any observations found, and any necessary action/s recommended.

Performance Rating

CDX Tasksheet Number: C743 **2008 NATEF Reference Number: 2A9**

☐ 0 ☐ 1 ☐ 2 ☐ 3 ☐ 4

Supervisor/instructor signature ______________________ Date______

In-Vehicle Transmission/Transaxle Maintenance and Repair

Student/intern information:

Name________________________ Date__________ Class________________

Vehicle used for this activity:

Year__________ Make________________________ Model______________

Odometer________________________ VIN ________________________

Time off________

Time on________

Total time________

Learning Objective/Task	CDX Tasksheet Number	2008 NATEF Reference Number	2008 NATEF Priority Level
• Inspect, adjust, and replace manual valve shift linkage, transmission range sensor/switch, and park/neutral position switch.	C683	2B1	P-2
• Inspect and replace external seals, gaskets, and bushings.	C748	2B2	P-2
• Diagnose electronic transmission control systems using a scan tool; determine necessary action.	C075	2B4	P-1
• Inspect, test, adjust, repair, or replace electrical/electronic components and circuits, including computers, solenoids, sensors, relays, terminals, connectors, switches, and harnesses.	C601	2B3	P-1
• Inspect, replace, and align powertrain mounts.	C602	2B5	P-2
• Service transmission; perform visual inspection; replace fluid and filters.	C603	2B6	P-1

Recommended Resource Materials

- CDX Automotive program
- Technical service bulletins, shop manuals, and any other information applicable to the specific vehicle or components you are working on
- Class notes

Materials Required

- Vehicle with automatic transmission/transaxle
- Vehicle hoist
- Scan tool
- DVOM/DMM
- Floor jack
- Drain pan
- Torque wrench
- Filter and fluid

Some Safety Issues to Consider

- Vehicle hoists are important tools that increase productivity and make the job easier. However, they can also cause severe injury or death if used improperly. Make sure you follow the manufacturer's operation procedures. Also, make sure you have your supervisor/instructor's permission to use any particular type of lifting equipment.
- When running any vehicles in the shop, make sure you use the shop's exhaust ventilation system to discharge all exhaust gas safely outside.
- Diagnosis of this fault may require test driving the vehicle on the school grounds. Attempt this task only with full permission from your instructor and follow all the guidelines exactly.
- This procedure may require cleaning the transmission or axle assembly prior to inspection. If so, use the proper equipment and procedures to carry this out safely. Also, get your supervisor/instructor's permission prior to cleaning.
- Always wear the correct protective eyewear and clothing and use the appropriate safety equipment, as well as fender covers, seat protectors, and floor mat protectors.
- Make sure you understand and observe all legislative and personal safety procedures when carrying out practical assignments. If you are unsure of what these are, ask your supervisor/instructor.

Performance Standard

0–No exposure: No information or practice provided during the program; complete training required
1–Exposure only: General information provided with no practice time; close supervision needed; additional training required
2–Limited practice: Has practiced job during training program; additional training required to develop skill
3–Moderately skilled: Has performed job independently during training program; limited additional training may be required
4–Skilled: Can perform job independently with no additional training

Time off_______________

Time on_______________

Total time_______________

▶ TASK Inspect, adjust, and replace manual valve shift linkage, transmission range sensor/switch, and park/neutral position switch. **C683 2B1**

1. Research the specifications and procedures for inspecting, adjusting, and replacing the manual valve shift linkage, transmission range sensor/switch, and park/neutral position switch in the appropriate service information.
   a. List, or print off and attach to this sheet, the procedure for inspecting, adjusting, and replacing these components:

   b. List any specified precautions:

2. Inspect the following components for proper operation and list your observations.
   a. Manual valve shift linkage:

   b. Transmission range sensor/switch:

   c. Park/neutral position switch:

3. Ask your instructor which components you should replace.
   a. Manual valve shift linkage: **Yes/No** (Circle one)
   b. Transmission range sensor/switch: **Yes/No** (Circle one)
   c. Park/neutral position switch: **Yes/No** (Circle one)
4. Remove the appropriate component/s following the specified procedure. List your observation/s:

5. Have your supervisor/instructor verify removal and initial below.
   a. Supervisor/instructor's initials: ______________________________
6. Replace each component following the specified procedure.
7. Adjust each component following the specified procedure.
8. Have your supervisor/instructor verify satisfactory completion of this procedure, any observations found, and any necessary action/s recommended.

Performance Rating

CDX Tasksheet Number: C683 **2008 NATEF Reference Number: 2B1**

☐	☐	☐	☐	☐
0	1	2	3	4

Supervisor/instructor signature ______________________________ Date__________

▸ TASK Inspect and replace external seals, gaskets, and bushings. C748 2B2

Vehicle used for this activity:

Year__________ Make______________________ Model______________

Odometer______________________ VIN ______________________

Time off__________

Time on__________

Total time__________

1. Safely raise the vehicle on a hoist.
2. Inspect the transmission/transaxle seals and gaskets for leaks.
   a. Identify and list the source of any leak/s:

3. Research the specifications and procedure for replacing the faulty component.
   a. List, or print off and attach to this sheet, the procedure for replacing the component:

b. List any torque specifications for this task:

c. List the flat rate time to replace the seal, gasket, or bushing: ____________________ hr

4. Have your supervisor/instructor approve the replacement of the failed component and initial below.
   a. Supervisor/instructor's initials: ______________________________
5. Remove the component following the specified procedures. List your observation/s:

6. Have your supervisor/instructor verify removal and your observations and initial below.
   a. Supervisor/instructor's initials: ______________________________
7. Install a new seal, gasket, and/or bushing. List your observation/s:

8. Have your supervisor/instructor verify satisfactory completion of this procedure, any observations found, and any necessary action/s recommended.

Performance Rating

CDX Tasksheet Number: C748 **2008 NATEF Reference Number: 2B2**

☐ 0 ☐ 1 ☐ 2 ☐ 3 ☐ 4

Supervisor/instructor signature ____________________ Date____________

▶ TASK Diagnose electronic transmission control systems using a scan tool; determine necessary action.

C075 2B4

Vehicle used for this activity:

Year____________ Make______________________ Model______________

Odometer______________________ VIN ______________________

Time off____________

Time on____________

Total time____________

1. List the customer concern related to an electronic transmission control system issue:

2. Connect an appropriate scan tool to the data link connector (DLC). Retrieve and list any transmission-related codes:

3. Research the specifications and procedures for diagnosing the code/s.
   a. List the possible cause/s of the concern:

   b. List, or print off and attach to this sheet, the procedure for diagnosing the concern:

4. Perform the diagnostic tests following the specified procedure using any appropriate test equipment.
   a. List your tests and test results:

5. Determine any necessary action/s to correct the fault:

6. Have your supervisor/instructor verify satisfactory completion of this procedure, any observations found, and any necessary action/s recommended.

Performance Rating

CDX Tasksheet Number: C075 **2008 NATEF Reference Number: 2B4**

☐ 0 ☐ 1 ☐ 2 ☐ 3 ☐ 4

Supervisor/instructor signature ________________ Date________

Total time______________

Time on______________

Time off______________

▶ TASK Inspect, test, adjust, repair, or replace electrical/electronic components and circuits, including computers, solenoids, sensors, relays, terminals, connectors, switches, and harnesses.

C601 2B3

1. Research the specifications and procedures for inspecting, testing, adjusting, and replacing the electrical/electronic components and sensors.
   a. List, or print off and attach to this sheet, the procedure for inspecting, testing, adjusting, and replacing these components:

   b. List any specified precautions:

2. Inspect the following components for proper operation and list your observations.
   a. Computer:

   b. Solenoids/relays:

   c. Sensors/switches:

   d. Terminals/connectors/harnesses:

3. Ask your instructor which components you should replace.
    a. Computer: **Yes/No** (Circle one)
    b. Solenoids/relays: **Yes/No** (Circle one)
    c. Sensors/switches: **Yes/No** (Circle one)
    d. Terminals/connectors/harnesses: **Yes/No** (Circle one)
4. Remove the appropriate component/s following the specified procedure. List your observation/s:

5. Have your supervisor/instructor verify removal and your observations and initial below.
    a. Supervisor/instructor's initials: ______________________________
6. Replace each component following the specified procedure.
7. Adjust each component following the specified procedure.
8. Have your supervisor/instructor verify satisfactory completion of this procedure, any observations found, and any necessary action/s recommended.

Performance Rating

CDX Tasksheet Number: C601 **2008 NATEF Reference Number: 2B3**

☐ 0 ☐ 1 ☐ 2 ☐ 3 ☐ 4

Supervisor/instructor signature ______________________________ Date______________

▶ TASK Inspect, replace, and align powertrain mounts. **C602 2B5**

Vehicle used for this activity:

Year____________ Make______________________________ Model____________________

Odometer__________________________ VIN ______________________________________

Time off________

Time on________

Total time________

1. Research the procedure to inspect and replace the powertrain mounts in the appropriate service information.
    a. List the type of mount/s this vehicle uses: ______________________________
    b. List, or print off and attach to this sheet, the procedure for inspecting the mount/s:

    c. List, or print off and attach to this sheet, the precautions for performing this task:

2. Following the specified procedure, inspect the powertrain mounts and list your observations:

3. Determine any necessary action/s:

4. Have your supervisor/instructor verify your answers and initial below.
   a. Supervisor/instructor's initials: ______________________________

5. With your supervisor/instructor's permission, remove one or more powertrain mounts following the specified procedure. Inspect the mount/s and list your observations:

6. Have your supervisor/instructor verify removal of the mount and your recommendations and initial below.
   a. Supervisor/instructor's initials: ______________________________

7. Reinstall the powertrain mount/s according to the specified procedure.

8. Determine any necessary action/s:

9. Have your supervisor/instructor verify satisfactory completion of this procedure, any observations found, and any necessary action/s recommended.

Performance Rating

CDX Tasksheet Number: C602 **2008 NATEF Reference Number: 2B5**

☐ 0 ☐ 1 ☐ 2 ☐ 3 ☐ 4

Supervisor/instructor signature ______________________________ Date______________

▶ TASK Service transmission; perform visual inspection; replace fluid and filters. **C603 2B6**

Time off________

Time on________

Total time________

Vehicle used for this activity:

Year____________ Make______________________ Model______________________

Odometer______________________ VIN ______________________

1. Research the specifications and procedure in the appropriate service information.
   a. Transmission service interval: ______________________ mi/km
   b. Transmission fluid type: ______________________
   c. Transmission fluid capacity: ______________________ qt/lt
   d. Pan bolts torque: ______________________ ft-lb/Nm
   e. Filter screw torque, if needed: ______________________ ft-lb/Nm
   f. List the steps required to properly check the fluid level:

   g. List, or print off and attach to this sheet, the procedure for servicing the transmission:

2. Following the specified procedure, remove the transmission pan and filter. Inspect any residue and debris in the bottom of the pan or stuck to the filter. List your observations:

3. Determine any necessary action/s:

4. Have your supervisor/instructor verify your answers and removal of the pan and filter.
   a. Supervisor/instructor's initials: ______________________
5. Following the specified procedure, install the filter, pan, initial amount of fluid, and any other removed components.

NOTE ▸ Make sure you fill the transmission with the proper amount of the specified fluid. Most service information lists two fluid capacities: one for a drain and refill and the second for an overhaul. Make sure you use the proper specification.

6. Follow the specified procedure to circulate the fluid and bleed any air from the system. This usually involves moving the gear selector through each position and then checking the fluid level. Add fluid to the proper level.
7. Verify the correct operation of the transmission. This may require a test drive. Get your supervisor/instructor's permission before performing this step, and list your observation/s:

8. Have your supervisor/instructor verify satisfactory completion of this procedure, any observations found, and any necessary action/s recommended.

Performance Rating

CDX Tasksheet Number: C603 **2008 NATEF Reference Number: 2B6**

☐	☐	☐	☐	☐
0	1	2	3	4

Supervisor/instructor signature ______________________ Date______

Transmission Removal and Replacement

Student/intern information:

Name________________ Date________ Class__________

Vehicle used for this activity:

Year________ Make______________ Model__________

Odometer______________ VIN ______________

Time off________

Time on________

Total time________

Learning Objective/Task	CDX Tasksheet Number	2008 NATEF Reference Number	2008 NATEF Priority Level
• Remove and reinstall transmission/transaxle and torque converter; inspect engine core plugs, rear crankshaft seal, dowel pins, dowel pin holes, and mating surfaces.	C604	2C1	P-1
• Inspect converter flex (drive) plate, converter attaching bolts, converter pilot, converter pump drive surfaces, converter end play, and crankshaft pilot bore.	C605	2C7	P-2
• Inspect, leak test, and flush or replace transmission/transaxle oil cooler, lines, and fittings.	C749	2C6	P-1
• Measure transmission/transaxle end play or preload; determine necessary action.	C689	2C10	P-1

Recommended Resource Materials

- CDX Automotive program
- Technical service bulletins, shop manuals, and any other information applicable to the specific vehicle or components you are working on
- Class notes

Materials Required

- Vehicle with automatic transmission/transaxle
- Vehicle hoist
- Dial indicator
- Drain pan
- Cooler flushing equipment

Some Safety Issues to Consider

- Vehicle hoists are important tools that increase productivity and make the job easier. However, they can also cause severe injury or death if used improperly. Make sure you follow the manufacturer's operation procedures. Also, make sure you have your supervisor/instructor's permission to use any particular type of lifting equipment.
- When running any vehicles in the shop, make sure you use the shop's exhaust ventilation system to discharge all exhaust gas safely outside.
- Diagnosis of this fault may require test driving the vehicle on the school grounds. Attempt this task only with full permission from your instructor and follow all the guidelines exactly.
- This procedure may require cleaning the transmission or axle assembly prior to inspection. If so, use the proper equipment and procedures to carry this out safely. Also, get your supervisor/instructor's permission prior to cleaning.
- Always wear the correct protective eyewear and clothing and use the appropriate safety equipment, as well as fender covers, seat protectors, and floor mat protectors.
- Make sure you understand and observe all legislative and personal safety procedures when carrying out practical assignments. If you are unsure of what these are, ask your supervisor/instructor.

Performance Standard

0–No exposure: No information or practice provided during the program; complete training required
1–Exposure only: General information provided with no practice time; close supervision needed; additional training required
2–Limited practice: Has practiced job during training program; additional training required to develop skill
3–Moderately skilled: Has performed job independently during training program; limited additional training may be required
4–Skilled: Can perform job independently with no additional training

Total time ______

Time on ______

Time off ______

▶ TASK Remove and reinstall transmission/transaxle and torque converter; inspect engine core plugs, rear crankshaft seal, dowel pins, dowel pin holes, and mating surfaces.

C604 2C1

1. Research the following torque specifications and transmission/transaxle removal and installation procedure in the appropriate service information.
   a. Flex plate-to-crankshaft bolt torque: ______ ft-lb/Nm
   b. Torque converter-to-flex plate bolt torque: ______ ft-lb/Nm
   c. Transmission-to-block bolt torque: ______ ft-lb/Nm
   d. Transmission fluid type: ______
   e. Transmission fluid capacity: ______ qt/lt
   f. List, or print off and attach to this sheet, the procedure for removing and installing the transmission/transaxle:

2. Following the specified procedure, remove the transmission/transaxle from the vehicle.
3. Inspect the following components and list your observations.
   a. Engine core plugs:

   b. Rear crankshaft seal:

   c. Dowel pins and holes:

   d. Engine and bell housing mating surfaces:

4. Have your supervisor/instructor verify your observations and initial below.
   a. Supervisor/instructor's initials: ______________________________

NOTE ▸ At this time, you may want to overhaul the transmission/transaxle or perform other tasks such as inspecting/replacing the torque converter or flushing the transmission/transaxle cooler lines. Once any related tasks are completed and it is time to reinstall the transmission/transaxle, return to this page to complete the task and have it signed off.

5. Prepare the transmission to be reinstalled. Ensure that the torque converter is fully engaged in the transmission. Failure to do so may cause severe damage to the transmission/transaxle and the converter.
6. Position all wires, hoses, and tubes out of the way and in their specified positions. Secure them temporarily if necessary.

NOTE ▸ It is easy for wires and hoses to fall down and get pinched between the engine and bell housing. This will cause electrical issues in the vehicle as well as alignment issues with the transmission.

7. Following the specified procedure, reinstall the transmission/transaxle. Be sure to tighten all fasteners to the proper torque.
8. Refill the transmission/transaxle to the proper level.
9. Check that all cooler lines, electrical wires, linkages, and other removed components are properly installed and adjusted.
10. Have your supervisor/instructor verify that the vehicle is ready to be started and initial below.
    a. Supervisor/instructor's initials: ______________________________
11. Place exhaust hose/s over the exhaust pipe/s and set the parking brake.
12. Start the vehicle and move the transmission selector through each gear. Check the transmission fluid level and top off as necessary.
13. With your supervisor/instructor's permission, test drive the vehicle to ensure the transmission operates correctly. List your observations:

14. Double check your work and verify that there are no leaks. List your observations:

15. Have your supervisor/instructor verify satisfactory completion of this procedure, any observations found, and any necessary action/s recommended.

Performance Rating

CDX Tasksheet Number: C604 **2008 NATEF Reference Number: 2C1**

☐	☐	☐	☐	☐
0	1	2	3	4

Supervisor/instructor signature ______________________________ Date______________

Total time__________

Time on__________

Time off__________

▸ TASK Inspect converter flex (drive) plate, converter attaching bolts, converter pilot, converter pump drive surfaces, converter end play, and crankshaft pilot bore.

C605 2C7

1. Research the procedure and specifications for inspecting the torque converter, flex plate, and crankshaft pilot bore.
   a. Torque converter-to-flex plate bolt torque: ____________________ ft-lb/Nm
   b. Flex plate-to-crankshaft bolt torque: ____________________ ft-lb/Nm
   c. Converter end play: ____________________ in/mm
   d. Converter pilot diameter, if specified: ____________________ in/mm
   e. Crankshaft pilot bore, if specified: ____________________ in/mm
2. Following the specified procedure, inspect each of the listed items below and list your observations.
   a. Flex plate (cracks, warpage, elongated holes, starter ring gear wear, etc.):

   b. Converter attaching bolts (worn, stretched, rounded heads, etc.):

   c. Converter pilot (gouges, excessive rust, etc.):

   d. Crankshaft pilot bore (gouges, excessive rust, etc.):

   e. Converter pump drive surface (wear, nicks, burrs, etc.):

   f. Converter end play: ____________________ in/mm

3. Determine any necessary action/s:

4. Have your supervisor/instructor verify satisfactory completion of this procedure, any observations found, and any necessary action/s recommended.

Performance Rating

CDX Tasksheet Number: C605 **2008 NATEF Reference Number: 2C7**

☐ 0 ☐ 1 ☐ 2 ☐ 3 ☐ 4

Supervisor/instructor signature ______________________________ Date________

▶ TASK Inspect, leak test, and flush or replace transmission/transaxle oil cooler, lines, and fittings. C749 2C6

Time off________

Time on________

Total time________

1. Research the procedure to inspect, test, flush, and replace the cooler lines and fittings.
   a. List, or print off and attach to this sheet, the steps to test, flush, and replace the cooler lines:

2. Following the specified procedure, perform the tasks and list your observations.
   a. Inspect the cooler, lines, and fittings for damage or wear:

   b. Test the cooler, lines, and fittings for leaks:

   c. Flush or replace the cooler and lines:

3. Have your supervisor/instructor verify satisfactory completion of this procedure, any observations found, and any necessary action/s recommended.

Performance Rating

CDX Tasksheet Number: C749 **2008 NATEF Reference Number: 2C6**

☐ 0 ☐ 1 ☐ 2 ☐ 3 ☐ 4

Supervisor/instructor signature ______________________ Date______

Total time______

Time on______

Time off______

TASK Measure transmission/transaxle end play or preload; determine necessary action.

C689 2C10

1. Research the procedure and specifications to measure the end play or preload.
   a. Specified end play or preload: ______________________
   b. List, or print off and attach to this sheet, the steps to perform the end play or preload measurement:
   c. How is the end play or preload adjusted on this transmission/transaxle?
2. Following the specified procedure, measure the end play or preload: ______________________
3. Determine any necessary action/s:
4. Have your supervisor/instructor verify satisfactory completion of this procedure, any observations found, and any necessary action/s recommended.

Performance Rating

CDX Tasksheet Number: C689 **2008 NATEF Reference Number: 2C10**

☐ 0 ☐ 1 ☐ 2 ☐ 3 ☐ 4

Supervisor/instructor signature ______________________ Date______

Transmission Disassembly and Inspection

Student/intern information:

Name________________ Date________ Class____________

Vehicle used for this activity:

Year________ Make________________ Model____________

Odometer________________ VIN ________________

Time off________

Time on________

Total time________

Learning Objective/Task	CDX Tasksheet Number	2008 NATEF Reference Number	2008 NATEF Priority Level
• Disassemble, clean, and inspect transmission/transaxle.	C684	2C2	P-1
• Inspect, measure, and reseal oil pump assembly and components.	C688	2C9	P-1
• Inspect and measure planetary gear assembly components; determine necessary action.	C752	2C14	P-2
• Inspect bushings; determine necessary action.	C751	2C13	P-2
• Inspect, measure, and replace thrust washers and bearings.	C750	2C11	P-2
• Inspect roller and sprag clutch, races, rollers, sprags, springs, cages, and retainers; determine necessary action.	C759	2C21	P-1
• Inspect oil delivery circuits, including seal rings, ring grooves, and sealing surface areas, feed pipes, orifices, and check valves/balls.	C690	2C12	P-2
• Inspect case bores, passages, bushings, vents, and mating surfaces; determine necessary action.	C753	2C15	P-2
• Inspect servo and accumulator bores, pistons, seals, pins, springs, and retainers; determine necessary action.	C686	2C4	P-2
• Inspect bands and drums; determine necessary action.	C760	2C22	P-2
• Inspect clutch drum, piston, check balls, springs, retainers, seals, and friction and pressure plates; determine necessary action.	C756	2C18	P-2
• Measure clutch pack clearance; determine necessary action.	C757	2C19	P-1
• Air test operation of clutch and servo assemblies.	C758	2C20	P-1
• Inspect, measure, clean, and replace valve body (includes surfaces, bores, springs, valves, sleeves, retainers, brackets, check valves/balls, screens, spacers, and gaskets).	C685	2C3	P-2
• Inspect transaxle drive, link chains, sprockets, gears, bearings, and bushings; perform necessary action.	C754	2C16	P-2
• Inspect, measure, repair, adjust, or replace transaxle final drive components.	C755	2C17	P-2

Recommended Resource Materials

- CDX Automotive program
- Technical service bulletins, shop manuals, and any other information applicable to the specific vehicle or components you are working on
- Class notes

Materials Required

- Transmission/transaxle
- Dial indicator
- Feeler blades
- Straightedge
- Approved air nozzle
- Manufacturer-specific tools

Some Safety Issues to Consider

- Castings and machined parts may have sharp edges. Handle these parts with care.
- Many transmission parts are round and can easily roll off of tables and carts. Be sure to place them in a secure manner to prevent component damage or personal injury from falling objects.
- Compressed air can be very dangerous if applied to broken skin or body cavities. Always use approved air nozzles, and never point them at yourself or another person.
- Transmissions, transaxles, and their component assemblies are heavy and can slip easily. Work safely, get help, and do not overextend your abilities.
- Always wear the correct protective eyewear and clothing and use the appropriate safety equipment, as well as fender covers, seat protectors, and floor mat protectors.
- Make sure you understand and observe all legislative and personal safety procedures when carrying out practical assignments. If you are unsure of what these are, ask your supervisor/instructor.

Performance Standard

0–No exposure: No information or practice provided during the program; complete training required
1–Exposure only: General information provided with no practice time; close supervision needed; additional training required
2–Limited practice: Has practiced job during training program; additional training required to develop skill
3–Moderately skilled: Has performed job independently during training program; limited additional training may be required
4–Skilled: Can perform job independently with no additional training

Total time________

Time on________

Time off________

▶ TASK Disassemble, clean, and inspect transmission/transaxle. **C684 2C2**

1. Research the disassembly and inspection procedures in the appropriate service information.
   a. List any precautions for performing this task:

   b. List any special tools required to perform this task:

2. Following the specified procedure, disassemble the transmission/transaxle into its unit components, ensuring that you note their location and orientation. Make notes or take pictures as appropriate. Also, lay them out in a logical manner to facilitate reassembly.

> **NOTE ▶** Round objects such as clutch drums, ball bearings, roller bearings, and bushings can roll off of the work table if not secured, causing personal injury. This could also lead to damage or loss of the components. Always lay round objects on their side when possible, or store them in a work tray to prevent rolling.

3. Clean the components according to the specified procedure. Make sure you relocate them in the same logical sequence as before.

4. Perform a preliminary visual inspection of the components. This is an initial inspection designed to identify major issues with the transmission/transaxle. Further inspection of the subunits will occur later. List your observations:

5. Have your supervisor/instructor verify satisfactory completion of this procedure, any observations found, and any necessary action/s recommended.

Performance Rating

CDX Tasksheet Number: C684 **2008 NATEF Reference Number: 2C2**

☐	☐	☐	☐	☐
0	1	2	3	4

Supervisor/instructor signature ______________________ Date__________

▶ TASK Inspect, measure, and reseal oil pump assembly and components. C688 2C9

Time off__________

Time on__________

Total time__________

1. Research the procedure for disassembling and inspecting the oil pump in the appropriate service information.
   a. Type of pump (gerotor, crescent, vane, etc.): ______________________
   b. List, or print off and attach to this sheet, the clearance specifications for this pump:

   c. Pump housing bolt torque: ______________________ ft-lb/Nm
   d. Pump-to-case bolt torque: ______________________ ft-lb/Nm
2. Following the specified procedure, disassemble the oil pump. Be sure not to damage any of the surfaces or parts.

NOTE ▶ You may need to scribe a mark on the pump halves so that you can reassemble it in the proper orientation. Refer to the service information to see if this is the case.

3. Visually inspect the pump components, housing, and seal for damage or wear. List your observations:

4. Measure the specified clearances. List your readings:

5. Determine any necessary action/s:

6. Have your supervisor/instructor verify disassembly and your observations and initial below.
   a. Supervisor/instructor's initials: ______________________________
7. Following the specified procedure, reassemble the oil pump. Be sure to prelube the oil pump and replace any seals as necessary. List your observations:

8. Have your supervisor/instructor verify satisfactory completion of this procedure, any observations found, and any necessary action/s recommended.

Performance Rating

CDX Tasksheet Number: C688 **2008 NATEF Reference Number: 2C9**

☐ 0 ☐ 1 ☐ 2 ☐ 3 ☐ 4

Supervisor/instructor signature ______________________________ Date______________

Total time______________

Time on______________

Time off______________

▶ TASK Inspect and measure planetary gear assembly components; determine necessary action. **C752 2C14**

1. Research the procedure for inspecting and measuring the planetary gear/s in the appropriate service information.
   a. Type of planetary gear (Simpson, Ravigneaux, etc.): ______________________
   b. List, or print off and attach to this sheet, the clearance and other specifications for the planetary gear/s:

2. Following the specified procedure, perform any planetary gear assembly measurements. List your observations:

3. Visually inspect the planetary gear assemblies (including the sun gear, ring gear, planetary gears, thrust washers, and carrier assembly) for wear or damage. List your observations:

4. Determine any necessary action/s:

5. Have your supervisor/instructor verify satisfactory completion of this procedure, any observations found, and any necessary action/s recommended.

Performance Rating

CDX Tasksheet Number: C752 **2008 NATEF Reference Number: 2C14**

☐	☐	☐	☐	☐
0	1	2	3	4

Supervisor/instructor signature ____________________ Date__________

▸ TASK Inspect bushings; determine necessary action. C751 2C13

Time off__________

Time on__________

Total time__________

1. Research the procedure and any specifications for inspecting the bushings in the appropriate service information.
   a. List the name and any specifications for each bushing:

2. Following the specified procedure, inspect and/or measure each bushing. List your observations:

3. Determine any necessary action/s:

4. Have your supervisor/instructor verify satisfactory completion of this procedure, any observations found, and any necessary action/s recommended.

Performance Rating

CDX Tasksheet Number: C751 2008 NATEF Reference Number: 2C13

☐ 0 ☐ 1 ☐ 2 ☐ 3 ☐ 4

Supervisor/instructor signature ____________________ Date____________

Total time____________

Time on____________

Time off____________

TASK Inspect, measure, and replace thrust washers and bearings. C750 2C11

1. Research the procedure and specifications for inspecting, measuring, and replacing thrust washers and bearings in the appropriate service information.
   a. List the name and specified thickness for each thrust washer and thrust bearing:

   b. Which thrust washer/s or thrust bearing/s is the selective fit for obtaining the proper gear train end play? ____________________

2. Following the specified procedure, inspect and/or measure each thrust washer and thrust bearing. List your observations:

3. Determine any necessary action/s:

4. Have your supervisor/instructor verify satisfactory completion of this procedure, any observations found, and any necessary action/s recommended.

Performance Rating

CDX Tasksheet Number: C750 2008 NATEF Reference Number: 2C11

☐ 0 ☐ 1 ☐ 2 ☐ 3 ☐ 4

Supervisor/instructor signature ______ Date______

▶ TASK Inspect roller and sprag clutch, races, rollers, sprags, springs, cages, and retainers; determine necessary action. C759 2C21

Time off______

Time on______

Total time______

1. Research the procedure and any specifications for inspecting the roller or sprag clutch assemblies in the appropriate service information.
   a. List the device that each roller or sprag clutch holds:

   b. List, or print off and attach to this sheet, any inspection procedures for these components:

2. Following the specified procedure, inspect each roller and sprag clutch (including races, rollers, sprags, springs, cages, and retainers). List your observations:

3. Determine any necessary action/s:

4. Have your supervisor/instructor verify satisfactory completion of this procedure, any observations found, and any necessary action/s recommended.

Performance Rating

CDX Tasksheet Number: C759 2008 NATEF Reference Number: 2C21

☐ 0 ☐ 1 ☐ 2 ☐ 3 ☐ 4

Supervisor/instructor signature ______ Date______

Total time___________

Time on___________

Time off___________

▶ TASK Inspect oil delivery circuits, including seal rings, ring grooves, and sealing surface areas, feed pipes, orifices, and check valves/balls. C690 2C12

1. Research the procedure and any specifications for inspecting the oil delivery circuits, including seal rings, ring grooves, and sealing surface areas; feed pipes; orifices; and check valves/balls.
   a. List any specifications for the above components:

2. Following the specified procedure, inspect the oil delivery circuits (including the components above). List your observations:

3. Determine any necessary action/s:

4. Have your supervisor/instructor verify satisfactory completion of this procedure, any observations found, and any necessary action/s recommended.

Performance Rating

CDX Tasksheet Number: C690 **2008 NATEF Reference Number: 2C12**

☐ 0 ☐ 1 ☐ 2 ☐ 3 ☐ 4

Supervisor/instructor signature ___________________________ Date___________

Total time___________

Time on___________

Time off___________

▶ TASK Inspect case bores, passages, bushings, vents, and mating surfaces; determine necessary action. C753 2C15

1. Research the procedure and any specifications for inspecting case bores, passages, bushings, vents, and mating surfaces.
   a. List any specifications for the above components:

2. Following the specified procedure, inspect case bores, passages, bushings, vents, and mating surfaces. List your observations:

3. Determine any necessary action/s:

4. Have your supervisor/instructor verify satisfactory completion of this procedure, any observations found, and any necessary action/s recommended.

Performance Rating

CDX Tasksheet Number: C753 **2008 NATEF Reference Number: 2C15**

☐ 0 ☐ 1 ☐ 2 ☐ 3 ☐ 4

Supervisor/instructor signature ____________________ Date________

▶ TASK Inspect servo and accumulator bores, pistons, seals, pins, springs, and retainers; determine necessary action. C686 2C4

Time off________

Time on________

Total time________

1. Research the procedure and any specifications for inspecting servo and accumulator bores, pistons, seals, pins, springs, and retainers.
   a. List any specifications for the above components:

2. If not already completed, disassemble each servo and accumulator.
3. Following the specified procedure, inspect servo and accumulator bores, pistons, seals, pins, springs, and retainers. List your observations:

4. Determine any necessary action/s:

5. Have your supervisor/instructor verify satisfactory completion of this procedure, any observations found, and any necessary action/s recommended.

Performance Rating

CDX Tasksheet Number: C686 **2008 NATEF Reference Number: 2C4**

☐	☐	☐	☐	☐
0	1	2	3	4

Supervisor/instructor signature ____________________ Date ____________

Total time ____________

Time on ____________

Time off ____________

▶ TASK Inspect bands and drums; determine necessary action. C760 2C22

1. Research the procedure and specifications for inspecting the bands and drums.
   a. List what gear/s each band is applied in:

   b. List any specifications for each band and drum:

2. Following the specified procedure, inspect each band and drum for wear or damage. List your observations:

3. Determine any necessary action/s:

4. Have your supervisor/instructor verify satisfactory completion of this procedure, any observations found, and any necessary action/s recommended.

Performance Rating

CDX Tasksheet Number: C760 2008 NATEF Reference Number: 2C22

☐ 0 ☐ 1 ☐ 2 ☐ 3 ☐ 4

Supervisor/instructor signature ______________________ Date________

▶ TASK Inspect clutch drum, piston, check balls, springs, retainers, seals, and friction and pressure plates; determine necessary action. C756 2C18

Time off________

Time on________

Total time________

1. Research the procedure and specifications for inspecting the clutch drum, piston, check balls, springs, retainers, seals, and friction and pressure plates for each clutch.
   a. In the table below, list the name of each clutch this transmission/transaxle is equipped with and the number of steel plates and friction plates it contains.

Name of clutch	Number of steel plates	Number of friction plates

   b. List, or print off and attach to this sheet, any specifications related to the clutch drum, piston, check balls, springs, retainers, seals, and friction and pressure plates:

2. Following the specified procedure, disassemble and inspect the components for each clutch assembly. List your observations for each clutch:

3. Determine any necessary action/s:

4. Have your supervisor/instructor verify satisfactory completion of this procedure, any observations found, and any necessary action/s recommended.

Performance Rating

CDX Tasksheet Number: C756 **2008 NATEF Reference Number: 2C18**

☐ 0 ☐ 1 ☐ 2 ☐ 3 ☐ 4

Supervisor/instructor signature ______________________ Date__________

Total time________

Time on________

Time off________

▶ TASK Measure clutch pack clearance; determine necessary action.

C757 2C19

1. Research the procedure and specifications for measuring clutch pack clearance in the appropriate service information.
   a. List the specified clutch pack clearance for each clutch in the table below.

Name of clutch	Specified clearance (in/mm)

   b. List how the clutch pack clearance is adjusted on this transmission/transaxle:

2. If not already completed, reassemble each clutch following the specified procedure.
3. Following the specified procedure, measure the clutch pack clearance for each clutch and list your measurements in the table below.

Name of clutch	Measured clearance (in/mm)

4. Determine any necessary action/s:

5. Have your supervisor/instructor verify satisfactory completion of this procedure, any observations found, and any necessary action/s recommended.

Performance Rating

CDX Tasksheet Number: C757 **2008 NATEF Reference Number: 2C19**

☐ 0 ☐ 1 ☐ 2 ☐ 3 ☐ 4

Supervisor/instructor signature _______________ Date_______

▶ TASK Air test operation of clutch and servo assemblies. C758 2C20

Time off_______

Time on_______

Total time_______

NOTE ▶ Most clutch assemblies can usually be air tested outside of the transmission once they are completely assembled with the clutch pack. Therefore, now would be a good time to perform that task for each self-contained clutch pack. Servos, which operate bands, must usually be tested with the gear train and bands installed. You may have to come back to this task for supervisor/instructor sign-off when it is appropriate to air test any servos during the assembly of the transmission/transaxle.

1. Research the procedure for air testing each clutch and servo assembly.
   a. List any special tools needed for this procedure:

2. Following the specified procedure, air test each clutch assembly. List your observations for each one in the table below.

Name of clutch	Observations

3. Have your supervisor/instructor verify the above results and initial below.
   a. Supervisor/instructor's initials: _______________

4. Once the transmission/transaxle is assembled far enough to air test the servos, test each one following the specified procedure. List your observations in the table below.

Name of servo	Observations

5. Have your supervisor/instructor verify satisfactory completion of this procedure, any observations found, and any necessary action/s recommended.

Performance Rating

CDX Tasksheet Number: C758 **2008 NATEF Reference Number: 2C20**

☐ 0 ☐ 1 ☐ 2 ☐ 3 ☐ 4

Supervisor/instructor signature _______________ Date_______

Total time_______

Time on_______

Time off_______

▶ TASK Inspect, measure, clean, and replace valve body (includes surfaces, bores, springs, valves, sleeves, retainers, brackets, check valves/balls, screens, spacers, and gaskets). C685 2C3

1. Research the procedure and specifications to inspect, measure, clean, and replace the valve body.
   a. List, or print off and attach to this sheet, any precautions for this task:

   b. List, or print off and attach to this sheet, any specifications related to this task:

   c. List, or print off and attach to this sheet, the torque specifications for the valve body bolts, screws, and fasteners:

2. Following the specified procedure, disassemble, clean, measure, and inspect the valve body components. List your observations for each.
   a. Valve body and plate surfaces:

   b. Bores, valves, and sleeves:

   c. Springs:

   d. Retainers, spacers, and brackets:

   e. Check valves and check balls:

   f. Gaskets and screens:

3. Have your supervisor/instructor verify disassembly of the valve body and your observations and initial below.
   a. Supervisor/instructor's initials: ______________________________

4. Following the specified procedure, reassemble the valve body. Be careful to install all valves, springs, check balls, and gaskets in the proper orientation. Also, be sure to torque all fasteners in the correct sequence and to the proper torque.
5. Have your supervisor/instructor verify satisfactory completion of this procedure, any observations found, and any necessary action/s recommended.

Performance Rating

CDX Tasksheet Number: C685 **2008 NATEF Reference Number: 2C3**

❑ 0 ❑ 1 ❑ 2 ❑ 3 ❑ 4

Supervisor/instructor signature ______________________ Date__________

Total time__________

Time on__________

Time off__________

▶ TASK Inspect transaxle drive, link chains, sprockets, gears, bearings, and bushings; perform necessary action. C754 2C16

1. Research the procedure and specifications to inspect the transaxle drive, link chains, sprockets, gears, bearings, and bushings.
   a. List, or print off and attach to this sheet, any precautions for this task:

   b. List, or print off and attach to this sheet, any specifications related to this task:

2. Following the specified procedure, inspect the following components and list your observations.
   a. Link chains:

   b. Sprockets:

c. Gears:

d. Bearings:

e. Bushings:

3. Determine any necessary action/s:

4. Have your supervisor/instructor verify satisfactory completion of this procedure, any observations found, and any necessary action/s recommended.

Performance Rating

CDX Tasksheet Number: C754 **2008 NATEF Reference Number: 2C16**

☐ 0 ☐ 1 ☐ 2 ☐ 3 ☐ 4

Supervisor/instructor signature __________ Date__________

▶ TASK Inspect, measure, repair, adjust, or replace transaxle final drive components. C755 2C17

Time off__________

Time on__________

Total time__________

1. Research the procedure and specifications to inspect, measure, repair, adjust, or replace transaxle final drive components.
   a. List, or print off and attach to this sheet, any precautions for this task:

b. List any special tools required to complete this task:

c. List, or print off and attach to this sheet, any specifications related to this task:

2. Following the specified procedure, inspect, measure, repair, adjust, or replace the following components and list your observations.
   a. Link chains:

   b. Sprockets:

   c. Gears:

   d. Bearings:

   e. Bushings:

3. Determine any necessary action/s:

4. Have your supervisor/instructor verify satisfactory completion of this procedure, any observations found, and any necessary action/s recommended.

Performance Rating

CDX Tasksheet Number: C755 **2008 NATEF Reference Number: 2C17**

☐ 0 ☐ 1 ☐ 2 ☐ 3 ☐ 4

Supervisor/instructor signature ______________________ Date__________

Assemble Transmission

Student/intern information:

Name_______________ Date_______ Class_______________

Vehicle used for this activity:

Year_______ Make_______________ Model_______________

Odometer_______________ VIN_______________

Learning Objective/Task	CDX Tasksheet Number	2008 NATEF Reference Number	2008 NATEF Priority Level
• Assemble transmission/transaxle.	C083	2C5	P-1
• Install and seat torque converter to engage drive/splines.	C687	2C8	P-1

Time off_______

Time on_______

Total time_______

Recommended Resource Materials

- CDX Automotive program
- Technical service bulletins, shop manuals, and any other information applicable to the specific vehicle or components you are working on
- Class notes

Materials Required

- Transmission/transaxle
- Torque wrenches of appropriate size
- Feeler blades
- Approved air nozzle
- Specified assembly lubricant
- Manufacturer-specific tools

Some Safety Issues to Consider

- Castings and machined parts may have sharp edges. Handle these parts with care.
- Many transmission parts are round and can easily roll off of tables and carts. Be sure to place them in a secure manner to prevent component damage or personal injury from falling objects.
- Compressed air can be very dangerous if applied to broken skin or body cavities. Always use approved air nozzles, and never point them at yourself or another person.
- Transmissions, transaxles, and their component assemblies are heavy and can slip easily. Work safely, get help, and do not overextend your abilities.
- Always wear the correct protective eyewear and clothing and use the appropriate safety equipment, as well as fender covers, seat protectors, and floor mat protectors.
- Make sure you understand and observe all legislative and personal safety procedures when carrying out practical assignments. If you are unsure of what these are, ask your supervisor/instructor.

Performance Standard

0–No exposure: No information or practice provided during the program; complete training required
1–Exposure only: General information provided with no practice time; close supervision needed; additional training required
2–Limited practice: Has practiced job during training program; additional training required to develop skill
3–Moderately skilled: Has performed job independently during training program; limited additional training may be required
4–Skilled: Can perform job independently with no additional training

▶ TASK Assemble transmission/transaxle. C083 2C5

1. Research the reassembly procedure for this transmission/transaxle in the appropriate service information.
   a. List the specified assembly lubricant: _______________

Time off_______

Time on_______

Total time_______

b. List the special tools required to perform this task:

c. Transmission/transaxle end play: ______________________ in/mm

> **NOTE ▸** All of the component assemblies should already be properly reassembled. If they are not, reassemble them according to the specified procedures as listed in previous tasks. Ensure that they meet all specifications.

2. Following the specified procedure, reassemble the transmission, making sure all components are seated properly. Perform all specified checks along the way. List your observations:

3. Once the transmission/transaxle gear train and pump are fully installed, adjust any bands if necessary.
   a. Air check each servo and clutch. List each clutch and servo and your observations:

> **NOTE ▸** Your instructor may want to sign off the servo air check from task **C758: Air test operation of clutch and servo assemblies** on page 52 after you perform it here.

   b. Measure the gear train end play: ______________________ in/mm
4. Following the specified procedure, reinstall the valve body, linkage, filter, and pan/s.
5. Complete any other specified procedures related to the assembly process.
6. Have your supervisor/instructor verify satisfactory completion of this procedure, any observations found, and any necessary action/s recommended.

Performance Rating

CDX Tasksheet Number: C083 **2008 NATEF Reference Number: 2C5**

☐	☐	☐	☐	☐
0	1	2	3	4

Supervisor/instructor signature ______________________ Date______

▶ **TASK** Install and seat torque converter to engage drive/splines. **C687 2C8**

Time off________

Time on________

Total time________

1. Research the procedure for installing and seating the torque converter.
   a. Some transmissions list a specification and procedure for measuring the installed depth of the torque converter. If this is true for this transmission/transaxle, list, or print off and attach to this sheet, the procedure and specification for this procedure:

   b. Transmissions/transaxles without torque converter depth specifications can usually be installed by carefully counting the engagement of each component as the torque converter is installed. For example, on many transmissions, you would expect to feel the stator support, input shaft, and pump drive each engage as you install the converter. If you do not feel all three components engage, remove the converter, recheck the positioning of each item, and retry to install and seat the converter.
      i. List the components that need to engage for the converter to be fully seated:

2. Following the specified procedure, install and seat the torque converter. Perform any measurements or count the number of engagements as they occur. List your observations:

3. Have your supervisor/instructor verify satisfactory completion of this procedure, any observations found, and any necessary action/s recommended.

Performance Rating

CDX Tasksheet Number: C687 **2008 NATEF Reference Number: 2C8**

☐	☐	☐	☐	☐
0	1	2	3	4

Supervisor/instructor signature ________________ Date________

Appendix A: CDX/NATEF Correlation Guide for ASE 2: Automatic Transmission and Transaxle

A. General Transmission and Transaxle Diagnosis					
NATEF 2008 Reference and Priority	NATEF 2005 Reference and Priority	Tasksheet Title	2008 CDX Tasksheet Number	2005 CDX Tasksheet Number	Page(s)
2A1; P-1	2A01; P-1	Complete work order to include customer information, vehicle identifying information, customer concern, related service history, cause, and correction.	C738	C544	1-2, 5
2A2; P-1	2A02; P-1	Identify and interpret transmission/ transaxle concern; differentiate between engine performance and transmission/transaxle concerns; determine necessary action.	C599	C057	1, 4-5
2A3; P-1	2A03; P-1	Research applicable vehicle and service information, such as transmission/transaxle system operation, fluid type, vehicle service history, service precautions, and technical service bulletins.	C681	C058	1, 3
2A4; P-1	2A04; P-1	Locate and interpret vehicle and major component identification numbers.	C739	C059	1-3
2A5; P-1	2A05; P-1	Diagnose fluid loss and condition concerns; check fluid level in transmissions with and without dip-stick; determine necessary action.	C682	C060	11-13
2A6; P-1	2A06; P-1	Perform pressure tests (including transmissions/transaxles equipped with electronic pressure control); determine necessary action.	C740	C061	11, 13
2A7; P-3	2A07; P-3	Perform stall test; determine necessary action.	C741	C062	11, 14
2A8; P-3	2A08; P-3	Perform lock-up converter system tests; determine necessary action.	C063	C063	11, 15
N/A	2A09; P-2	REMOVED for 2008	N/A	C064	N/A
2A9; P-2	2A10; P-2	Diagnose noise and vibration concerns; determine necessary action.	C743	C065	17, 21-22
2A10; P-1	2A11; P-1	Diagnose transmission/transaxle gear reduction/multiplication concerns using driving, driven, and held member (power flow) principles.	C066	C066	17, 19-20
2A12; P-2	2A12; P-2	Diagnose pressure concerns in a transmission using hydraulic principles (Pascal's Law).	C747	C545	17, 20-21
N/A	2A13; P-2	REMOVED for 2008	N/A	C546	N/A
2A13; P-1	NEW	Diagnose electronic transmission/ transaxle control systems using appropriate test equipment and service information.	C600	NEW	17-19

B. In-Vehicle Transmission/Transaxle Maintenance and Repair					
NATEF 2008 Reference and Priority	NATEF 2005 Reference and Priority	Tasksheet Title	2008 CDX Tasksheet Number	2005 CDX Tasksheet Number	Page(s)
2B1; P-2	2B01; P-1	Inspect, adjust, and replace manual valve shift linkage, transmission range sensor/switch, and park/neutral position switch.	C683	C067	23-25
N/A	2C01 & 2C02; P-3	REMOVED for 2008	N/A	C069 & C070	N/A
2B2; P-2	2C03 & 2C04; P-3 & P-2	Inspect and replace external seals, gaskets, and bushings.	C748	C071 & C072	23, 25-26
2B3; P-1	NEW	Inspect, test, adjust, repair, or replace electrical/electronic components and circuits, including computers, solenoids, sensors, relays, terminals, connectors, switches, and harnesses.	C601	NEW	23, 28-29
N/A	2C05; P-2	REMOVED for 2008	N/A	C074	N/A
2B4; P-1	2C06; P-1	Diagnose electronic transmission control systems using a scan tool; determine necessary action.	C075	C075	23, 26-27
2B5; P-2	2C07; P-2	Inspect, replace, and align powertrain mounts.	C602	C076	23, 29-30
2B6; P-1	2B02; P-1	Service transmission; perform visual inspection; replace fluid and filters.	C603	C068	23, 31-32

C. Off-Vehicle Transmission and Transaxle Repair					
NATEF 2008 Reference and Priority	NATEF 2005 Reference and Priority	Tasksheet Title	2008 CDX Tasksheet Number	2005 CDX Tasksheet Number	Page(s)
2C1; P-1	2D101 & 2D102; P-2 & P-1	Remove and reinstall transmission/transaxle and torque converter; inspect engine core plugs, rear crankshaft seal, dowel pins, dowel pin holes, and mating surfaces.	C604	C077 & C078	33-35
2C2; P-1	2D103; P-1	Disassemble, clean, and inspect transmission/transaxle.	C684	C079	39-41
2C3; P-2	2D104; P-2	Inspect, measure, clean, and replace valve body (includes surfaces, bores, springs, valves, sleeves, retainers, brackets, check valves/balls, screens, spacers, and gaskets).	C685	C080	39, 52-54
2C4; P-2	2D105 & 2D106; P-3	Inspect servo and accumulator bores, pistons, seals, pins, springs, and retainers; determine necessary action.	C686	C081 & C082	39, 47-48
2C5; P-1	2D107; P-1	Assemble transmission/transaxle.	C083	C083	59-60
2C6; P-1	2D108; P-1	Inspect, leak test, and flush or replace transmission/transaxle oil cooler, lines, and fittings.	C749	C547	33, 37-38

2C7; P-2	2D201; P-2	Inspect converter flex (drive) plate, converter attaching bolts, converter pilot, converter pump drive surfaces, converter end play, and crankshaft pilot bore.	C605	C084	33, 36-37
N/A	2D202; P-2	REMOVED for 2008	N/A	C085	N/A
2C8; P-1	NEW	Install and seat torque converter to engage drive/splines.	C687	NEW	59, 61
2C9; P-1	2D203; P-1	Inspect, measure, and reseal oil pump assembly and components.	C688	C086	39, 41-42
2C10; P-1	2D301; P-1	Measure transmission/transaxle end play or preload; determine necessary action.	C689	C087	33, 38
2C11; P-2	2D302; P-2	Inspect, measure, and replace thrust washers and bearings.	C750	C088	39, 44-45
2C12; P-2	2D303; P-2	Inspect oil delivery circuits, including seal rings, ring grooves, and sealing surface areas, feed pipes, orifices, and check valves/balls.	C690	C089	39, 46
2C13; P-2	2D304; P-2	Inspect bushings; determine necessary action.	C751	C090	39, 43-44
2C14; P-2	2D305; P-2	Inspect and measure planetary gear assembly components; determine necessary action.	C752	C091	39, 42-43
2C15; P-2	2D306; P-2	Inspect case bores, passages, bushings, vents, and mating surfaces; determine necessary action.	C753	C092	39, 46-47
2C16; P-2	2D307; P-2	Inspect transaxle drive, link chains, sprockets, gears, bearings, and bushings; perform necessary action.	C754	C093	39, 54-55
2C17; P-2	2D308; P-2	Inspect, measure, repair, adjust, or replace transaxle final drive components.	C755	C094	39, 55-57
N/A	2D309; P-3	REMOVED for 2008	N/A	C095	N/A
2C18; P-2	2D401; P-2	Inspect clutch drum, piston, check balls, springs, retainers, seals, and friction and pressure plates; determine necessary action.	C756	C096	39, 49-50
2C19; P-1	2D402; P-1	Measure clutch pack clearance; determine necessary action.	C757	C097	39, 50-51
2C20; P-1	2D403; P-1	Air test operation of clutch and servo assemblies.	C758	C098	39, 51-52
2C21; P-1	2D404; P-1	Inspect roller and sprag clutch, races, rollers, sprags, springs, cages, and retainers; determine necessary action.	C759	C099	39, 45
2C22; P-2	2D405; P-2	Inspect bands and drums; determine necessary action.	C760	C100	39, 48-49
2C23; P-3	NEW	Describe the operational characteristics of a continuously variable transmission (CVT).	C606	NEW	1, 5-6
2C24; P-3	NEW	Describe the operational characteristics of a hybrid vehicle drive train.	C607	NEW	1, 6-7

Appendix B: Additional Automatic Transmission and Transaxle Tasksheets from 2005 NATEF Tasks

Student/intern information:

Name________________ Date________ Class________________

Vehicle used for this activity:

Year________ Make________________ Model________________

Odometer________________ VIN________________

Time off________

Time on________

Total time________

Learning Objective/Task	CDX Tasksheet Number	2005 NATEF Reference Number	2005 NATEF Priority Level
• Diagnose transmission/transaxle, mechanical, and vacuum control system concerns; determine necessary action.	C064	2A09	P-2
• Diagnose electrical/electronic concerns using principals of electricity (Ohm's Law).	C546	2A13	P-2

Recommended Resource Materials

- CDX Automotive program
- Technical service bulletins, shop manuals, and any other information applicable to the specific vehicle or components you are working on
- Class notes

Materials Required

- Depending on the type of concern, special diagnostic tools may be required. See your supervisor/instructor for instructions to identify what tools may be required.
- DVOM
- Vacuum gauge
- Vacuum pump

Some Safety Issues to Consider

- Use wheel chocks and ensure that the park brake is firmly applied to prevent the vehicle from moving when it is started.
- When running any vehicles in the shop, make sure you use the shop's exhaust ventilation system to discharge all exhaust gas safely outside.
- Only students who have their supervisor/instructor's direct permission should perform this task due to the safety concerns involved.
- Diagnosis of this fault may require test driving the vehicle on the school grounds. Attempt this task only with full permission from your supervisor/instructor and follow all the guidelines exactly.
- Always wear the correct protective eyewear and clothing and use the appropriate safety equipment, as well as fender covers, seat protectors, and floor mat protectors.
- Make sure you understand and observe all legislative and personal safety procedures when carrying out practical assignments. If you are unsure of what these are, ask your supervisor/instructor.

Performance Standard

0–No exposure: No information or practice provided during the program; complete training required

1–Exposure only: General information provided with no practice time; close supervision needed; additional training required

2–Limited practice: Has practiced job during training program; additional training required to develop skill

3–Moderately skilled: Has performed job independently during training program; limited additional training may be required

4–Skilled: Can perform job independently with no additional training

Total time__________

Time on__________

Time off__________

▸ TASK Diagnose transmission/transaxle, mechanical, and vacuum control system concerns; determine necessary action. **C064 2A09**

1. List the customer's concern:

2. If the vehicle is able to be test driven, obtain instructor permission to test drive the vehicle. Test drive the vehicle to verify the concern. List your observations:

3. Research the possible causes of the concern and the testing procedure in the appropriate service information.
   a. List the possible cause/s:

   b. List, or print off and attach to this sheet, the steps to diagnose the concern:

4. List your test/s and result/s:

5. With the information gained from the above activities, determine any necessary action/s to correct this fault:

6. Return the vehicle to its beginning condition and return any tools that you may have used to their proper locations.
7. Have your supervisor/instructor verify satisfactory completion of this procedure, any observations found, and any necessary action/s recommended.

Performance Rating

CDX Tasksheet Number: C064 **2005 NATEF Reference Number: 2A09**

☐ 0 ☐ 1 ☐ 2 ☐ 3 ☐ 4

Supervisor/instructor signature ____________________ Date__________

▶ TASK Diagnose electrical/electronic concerns using principals of electricity (Ohm's Law). C546 2A13

Time off______

Time on______

Total time______

1. List the electrical/electronic-related customer concern:

2. Research the particular complaint in the appropriate service manual and list the possible causes:

3. If necessary, inspect the transmission system using pressure testing equipment to determine the cause of the complaint. List your observations here:

4. Using a DVOM to test the electrical circuit in accordance with the service manual procedure and using Ohm's Law to confirm mathematically, inspect the system to determine the cause of the complaint. List your observations here:

5. Determine any necessary action/s to correct this fault:

6. Return the vehicle to its beginning condition and return any tools that you may have used to their proper locations.
7. Have your supervisor/instructor verify satisfactory completion of this procedure, any observations found, and any necessary action/s recommended.

Performance Rating

CDX Tasksheet Number: C546 **2005 NATEF Reference Number: 2A13**

☐ 0 ☐ 1 ☐ 2 ☐ 3 ☐ 4

Supervisor/instructor signature ______ Date______

Student/intern information:

Name__________ Date__________ Class__________

Vehicle used for this activity:

Year__________ Make__________ Model__________

Odometer__________ VIN __________

Total time__________

Time on__________

Time off__________

Learning Objective/Task	CDX Tasksheet Number	2005 NATEF Reference Number	2005 NATEF Priority Level
• Inspect, adjust, or replace (as applicable) vacuum modulator; inspect and repair or replace lines and hoses.	C069	2C01	P-3
• Inspect, repair, and replace governor assembly.	C070	2C02	P-3
• Inspect and replace speedometer drive gear, driven gear, vehicle speed sensor (VSS), and retainers.	C074	2C05	P-2

Recommended Resource Materials

- CDX Automotive program
- Technical service bulletins, shop manuals, and any other information applicable to the specific vehicle or components you are working on
- Class notes

Materials Required

- Vehicle or simulator with automatic transmission
- Vehicle hoist
- Vacuum gauge
- Vacuum pump
- Manufacturer-specific tools
- Basic hand tools

Some Safety Issues to Consider

- Vehicle hoists are important tools that increase productivity and make the job easier. However, they can also cause severe injury or death if used improperly. Make sure you follow the manufacturer's operation procedures. Also, make sure you have your supervisor/instructor's permission to use any particular type of lifting equipment.
- This procedure may require cleaning the transmission or axle assembly prior to inspection. If so, use the proper equipment and procedures to carry this out safely. Also, get your supervisor/instructor's permission prior to cleaning.
- Always wear the correct protective eyewear and clothing and use the appropriate safety equipment, as well as fender covers, seat protectors, and floor mat protectors.
- Make sure that you understand and observe all legislative and personal safety procedures when carrying out practical assignments. If you are unsure of what these are, check with your supervisor/instructor.

Performance Standard

0–No exposure: No information or practice provided during the program; complete training required
1–Exposure only: General information provided with no practice time; close supervision needed; additional training required
2–Limited practice: Has practiced job during training program; additional training required to develop skill
3–Moderately skilled: Has performed job independently during training program; limited additional training may be required
4–Skilled: Can perform job independently with no additional training

▶ TASK Inspect, adjust, or replace (as applicable) vacuum modulator; inspect and repair or replace lines and hoses.

C069 2C01

Time off________

Time on________

Total time________

1. Research the following specifications/procedures for this vehicle in the appropriate service manual.
   a. Transmission designation:
   b. List, or print off and attach to this sheet, the procedure to inspect, repair, and replace the vacuum modulator:
   c. List, or print off and attach to this sheet, the procedure to inspect and repair or replace fixed and flexible lines and hoses:
2. Safely raise the vehicle on a hoist.
3. Follow the service manual procedure for inspecting and adjusting or replacing the vacuum modulator. List your findings/actions:
4. Have your supervisor/instructor verify satisfactory completion and initial below.
   a. Supervisor/instructor's initials: ________________
5. Follow the service manual procedure for inspecting and repairing or replacing fixed and flexible vacuum lines and hoses, and list your findings/actions:
6. Return the vehicle to its beginning condition and return any tools that you may have used to their proper locations.

7. Have your supervisor/instructor verify satisfactory completion of this procedure, any observations found, and any necessary action/s recommended.

Performance Rating

CDX Tasksheet Number: C069 2005 NATEF Reference Number: 2C01

☐	☐	☐	☐	☐
0	1	2	3	4

Supervisor/instructor signature ____________________ Date________

Time off________

Time on________

Total time________

▶ TASK Inspect, repair, and replace governor assembly. C070 2C02

1. Research the following specifications/procedures for this vehicle in the appropriate service manual.
   a. Transmission designation:

   b. List, or print off and attach to this sheet, the procedure to inspect, repair, and replace the governor assembly:

2. Safely raise the vehicle on a hoist.
3. Follow the service manual procedure for inspecting, adjusting, or replacing the governor assembly, and list your findings/actions:

4. Return the vehicle to its beginning condition and return any tools that you may have used to their proper locations.
5. Have your supervisor/instructor verify satisfactory completion of this procedure, any observations found, and any necessary action/s recommended.

Performance Rating

CDX Tasksheet Number: C070 2005 NATEF Reference Number: 2C02

☐	☐	☐	☐	☐
0	1	2	3	4

Supervisor/instructor signature ____________________ Date________

▶ TASK Inspect and replace speedometer drive gear, driven gear, vehicle speed sensor (VSS), and retainers.

C074 2C05

Time off________

Time on________

Total time________

1. Research the following specifications/procedures for this vehicle in the appropriate service manual.
   a. Transmission designation:

   b. List, or print off and attach to this sheet, the procedure to inspect and replace the speedometer drive gear, driven gear, vehicle speed sensor (VSS), and retainers in the transmission:

2. Safely raise the vehicle on a hoist (if applicable).
3. Follow the service manual procedure to remove and inspect the speedometer drive gear, driven gear, vehicle speed sensor (VSS), and retainers in the transmission.
   a. List your findings and recommendations:

4. Have your supervisor/instructor verify satisfactory completion and initial below.
   a. Supervisor/instructor's initials: ________________
5. Follow the service manual procedure to replace the speedometer drive gear, driven gear, vehicle speed sensor (VSS), and retainers in the transmission.
   a. List your findings/actions:

6. Return the vehicle to its beginning condition and return any tools that you may have used to their proper locations.
7. Have your supervisor/instructor verify satisfactory completion of this procedure, any observations found, and any necessary action/s recommended.

Performance Rating

CDX Tasksheet Number: C074 **2005 NATEF Reference Number: 2C05**

☐	☐	☐	☐	☐
0	1	2	3	4

Supervisor/instructor signature ________________ Date________

Total time________

Time on________

Time off________

Student/intern information:

Name________________ Date________ Class________________

Vehicle used for this activity:

Year________ Make________________ Model________________

Odometer________________ VIN ________________

Learning Objective/Task	CDX Tasksheet Number	2005 NATEF Reference Number	2005 NATEF Priority Level
• Measure torque converter end play and check for interference; check stator clutch.	C085	2D202	P-2
• Inspect and reinstall parking pawl, shaft, spring, and retainer; determine necessary action.	C095	2D309	P-3

Recommended Resource Materials

- CDX Automotive program
- Technical service bulletins, shop manuals, and any other information applicable to the specific vehicle or components you are working on
- Class notes

Materials Required

- Automatic transmission/transaxle
- Supervisor/instructor-determined measurement devices
- Specified special OEM equipment

Some Safety Issues to Consider

- This procedure may require cleaning the transmission or axle assembly prior to inspection. If so, use the proper equipment and procedures to carry this out safely. Also, get your supervisor/instructor's permission prior to cleaning.
- Always wear the correct protective eyewear and clothing and use the appropriate safety equipment, as well as fender covers, seat protectors, and floor mat protectors.
- Make sure that you understand and observe all legislative and personal safety procedures when carrying out practical assignments. If you are unsure of what these are, check with your supervisor/instructor.

Performance Standard

0–No exposure: No information or practice provided during the program; complete training required
1–Exposure only: General information provided with no practice time; close supervision needed; additional training required
2–Limited practice: Has practiced job during training program; additional training required to develop skill
3–Moderately skilled: Has performed job independently during training program; limited additional training may be required
4–Skilled: Can perform job independently with no additional training

Total time________

Time on________

Time off________

▶ TASK Measure torque converter end play and check for interference; check stator clutch. **C085 2D202**

1. Research the procedure for measuring the torque converter end play for this particular transmission/transaxle in the appropriate service manual.
   a. List the transmission/transaxle torque converter end play: ________________ in/mm

2. Following the procedure outlined in the shop manual, examine the transmission and torque converter assemblies for obvious wear or damage.
   a. List your findings:

3. Following the shop manual procedure (and workshop manual diagrams for assistance), measure the torque converter end play and check for interference. Compare your findings to OEM specifications.
   a. Measured torque converter end play: ____________________ in/mm
   b. Is there any interference between the torque convertor turbine, impellor, and stator? **Yes/No** (Circle one)
   c. List your observations:

4. Following the shop manual procedure, check the stator clutch for one-way operation. List your observations:

5. Determine any necessary action/s:

6. Return the vehicle to its beginning condition and return any tools that you may have used to their proper locations.
7. Have your supervisor/instructor verify satisfactory completion of this procedure, any observations found, and any necessary action/s recommended.

Performance Rating

CDX Tasksheet Number: C085 **2005 NATEF Reference Number: 2D202**

☐ 0 ☐ 1 ☐ 2 ☐ 3 ☐ 4

Supervisor/instructor signature ____________________ Date____________

Total time____________

Time on____________

Time off____________

▶ TASK Inspect and reinstall parking pawl, shaft, spring, and retainer; determine necessary action.

C095 2D309

1. Research the procedure for inspecting and reinstalling the parking pawl for this particular transmission/transaxle in the appropriate service manual.
   a. List any specifications/adjustments related to this task:

2. Following the procedure outlined in the shop manual, disassemble the transmission sufficiently to enable the removal of the parking pawl mechanism. As you remove components, note their location and place them on a workbench in a logical manner to allow inspection of the components and reassembly of the unit to be carried out in an efficient manner.
3. Have your supervisor/instructor verify the satisfactory completion of this element of the task and initial below.
   a. Supervisor/instructor's initials: ____________________
4. Following the shop manual procedure (and workshop manual diagrams for assistance), inspect the parking pawl, shaft, spring, and retainer. List your observations:

5. Determine any necessary action/s:

6. Have your supervisor/instructor verify satisfactory completion of this element of the task and initial below.
   a. Supervisor/instructor's initials: ____________________
7. Replace any components as needed and then reassemble the parking pawl in its correct location.
8. Return the vehicle to its beginning condition and return any tools that you may have used to their proper locations.
9. Have your supervisor/instructor verify satisfactory completion of this procedure, any observations found, and any necessary action/s recommended.

Performance Rating

CDX Tasksheet Number: C095 **2005 NATEF Reference Number: 2D309**

☐ 0 ☐ 1 ☐ 2 ☐ 3 ☐ 4

Supervisor/instructor signature ____________________ Date__________